This Book Belongs To

Name : _____

Phone : _____

Address : _____

Project Name :

Foreman :

Project No :

Date :

Day :

Visitors

Schedule

Problems

Safety Issues

Summary Of Work

Signature :

Employee	Trade	Hours	Overtime

Equipment On Site	No. Units

Materials Delivered	No. Units	Equipment Rented	Rate

Others

Notes :

Project Name :

Foreman :

Project No :

Date :

Day :

Visitors	Schedule

Problems	Safety Issues

Summary Of Work

Signature :

Employee	Trade	Hours	Overtime

Equipment On Site	No. Units

Materials Delivered	No. Units	Equipment Rented	Rate

Others

Notes :

Project Name : _____

Foreman : _____

Project No :
Date :
Day :

Visitors	Schedule

Problems	Safety Issues

Summary Of Work

Signature : _____

Employee	Trade	Hours	Overtime

Equipment On Site	No. Units

Materials Delivered	No. Units	Equipment Rented	Rate

Others

Notes :

Project Name :

Foreman :

Project No :
Date :
Day :

Visitors	Schedule

Problems	Safety Issues

Summary Of Work

Signature :

Employee	Trade	Hours	Overtime

Equipment On Site	No. Units

Materials Delivered	No. Units	Equipment Rented	Rate

Others

Notes :

Project Name :

Foreman :

Project No :

Date :

Day :

Visitors	Schedule

Problems	Safety Issues

Summary Of Work

Signature :

Employee	Trade	Hours	Overtime

Equipment On Site	No. Units

Materials Delivered	No. Units	Equipment Rented	Rate

Others

Notes :

Project Name :

Foreman :

Project No :

Date :

Day :

Visitors	Schedule

Problems	Safety Issues

Summary Of Work

Signature :

Employee	Trade	Hours	Overtime

Equipment On Site	No. Units

Materials Delivered	No. Units	Equipment Rented	Rate

Others

Notes :

Project Name : _____

Foreman : _____

Project No :
Date :
Day :

Visitors	Schedule

Problems	Safety Issues

Summary Of Work

Signature : _____

Employee	Trade	Hours	Overtime

Equipment On Site	No. Units

Materials Delivered	No. Units	Equipment Rented	Rate

Others

Notes :

Project Name :

Foreman :

Project No :

Date :

Day :

Visitors	Schedule

Problems	Safety Issues

Summary Of Work

Signature :

Employee	Trade	Hours	Overtime

Equipment On Site	No. Units

Materials Delivered	No. Units	Equipment Rented	Rate

Others

Notes :

Project Name : _____

Foreman : _____

Project No :
Date :
Day :

Visitors

Schedule

Problems

Safety Issues

Summary Of Work

Signature : _____

Employee	Trade	Hours	Overtime

Equipment On Site	No. Units

Materials Delivered	No. Units	Equipment Rented	Rate

Others

Notes :

Project Name :

Foreman :

Project No :

Date :

Day :

Visitors

Schedule

Problems

Safety Issues

Summary Of Work

Signature :

Employee	Trade	Hours	Overtime

Equipment On Site	No. Units

Materials Delivered	No. Units	Equipment Rented	Rate

Others

Notes :

Project Name :	Project No :
	Date :
Foreman :	Day :

Visitors	Schedule

Problems	Safety Issues

Summary Of Work

Signature :

Employee	Trade	Hours	Overtime

Equipment On Site	No. Units

Materials Delivered	No. Units	Equipment Rented	Rate

Others

Notes :

Project Name :

Foreman :

Project No :

Date :

Day :

Visitors

Schedule

Problems

Safety Issues

Summary Of Work

Signature :

Employee	Trade	Hours	Overtime

Equipment On Site	No. Units

Materials Delivered	No. Units	Equipment Rented	Rate

Others

Notes :

Project Name :

Foreman :

Project No :
Date :
Day :

Visitors

Schedule

Problems

Safety Issues

Summary Of Work

Signature :

Employee	Trade	Hours	Overtime

Equipment On Site	No. Units

Materials Delivered	No. Units	Equipment Rented	Rate

Others

Notes :

Project Name : _____

Foreman : _____

Project No :
Date :
Day :

Visitors

Schedule

Problems

Safety Issues

Summary Of Work

Signature : _____ ✏

Employee	Trade	Hours	Overtime

Equipment On Site	No. Units

Materials Delivered	No. Units	Equipment Rented	Rate

Others

Notes :

Project Name : _____

Foreman : _____

Project No :
Date :
Day :

Visitors	Schedule

Problems	Safety Issues

Summary Of Work

Signature : _____

Employee	Trade	Hours	Overtime

Equipment On Site	No. Units

Materials Delivered	No. Units	Equipment Rented	Rate

Others

Notes :

Project Name :

Foreman :

Project No :
Date :
Day :

Visitors	Schedule

Problems	Safety Issues

Summary Of Work

Signature :

Employee	Trade	Hours	Overtime

Equipment On Site	No. Units

Materials Delivered	No. Units	Equipment Rented	Rate

Others

Notes :

Project Name : _____

Foreman : _____

Project No :
Date :
Day :

Visitors	Schedule

Problems	Safety Issues

Summary Of Work

Signature : _____

Employee	Trade	Hours	Overtime

Equipment On Site	No. Units

Materials Delivered	No. Units	Equipment Rented	Rate

Others

Notes :

Project Name : _____

Foreman : _____

Project No :
Date :
Day :

Visitors	Schedule

Problems	Safety Issues

Summary Of Work

Signature : _____

Employee	Trade	Hours	Overtime

Equipment On Site	No. Units

Materials Delivered	No. Units	Equipment Rented	Rate

Others

Notes :

Project Name :

Foreman :

Project No :

Date :

Day :

Visitors	Schedule

Problems	Safety Issues

Summary Of Work

Signature :

Employee	Trade	Hours	Overtime

Equipment On Site	No. Units

Materials Delivered	No. Units	Equipment Rented	Rate

Others

Notes :

Project Name :

Foreman :

Project No :

Date :

Day :

Visitors	Schedule

Problems	Safety Issues

Summary Of Work

Signature :

Employee	Trade	Hours	Overtime

Equipment On Site	No. Units

Materials Delivered	No. Units	Equipment Rented	Rate

Others

Notes :

Project Name : _____

Foreman : _____

Project No :
Date :
Day :

Visitors	Schedule

Problems	Safety Issues

Summary Of Work

Signature : _____

Employee	Trade	Hours	Overtime

Equipment On Site	No. Units

Materials Delivered	No. Units	Equipment Rented	Rate

Others

Notes :

Project Name :

Foreman :

Project No :
Date :
Day :

Visitors

Schedule

Problems

Safety Issues

Summary Of Work

Signature :

Employee	Trade	Hours	Overtime

Equipment On Site	No. Units

Materials Delivered	No. Units	Equipment Rented	Rate

Others

Notes :

Project Name :	Project No :
	Date :
Foreman :	Day :

Visitors	Schedule

Problems	Safety Issues

Summary Of Work

Signature :

Employee	Trade	Hours	Overtime

Equipment On Site	No. Units

Materials Delivered	No. Units	Equipment Rented	Rate

Others

Notes :

Project Name : _____

Foreman : _____

Project No :
Date :
Day :

Visitors	Schedule

Problems	Safety Issues

Summary Of Work

Signature : _____ ✎

Employee	Trade	Hours	Overtime

Equipment On Site	No. Units

Materials Delivered	No. Units	Equipment Rented	Rate

Others

Notes :

Project Name :

Foreman :

Project No :

Date :

Day :

Visitors	Schedule

Problems	Safety Issues

Summary Of Work

Signature :

Employee	Trade	Hours	Overtime

Equipment On Site	No. Units

Materials Delivered	No. Units	Equipment Rented	Rate

Others

Notes :

Project Name : _____

Foreman : _____

Project No :
Date :
Day :

Visitors

Schedule

Problems

Safety Issues

Summary Of Work

Signature : _____

Employee	Trade	Hours	Overtime

Equipment On Site	No. Units

Materials Delivered	No. Units	Equipment Rented	Rate

Others

Notes :

Project Name : _____

Foreman : _____

Project No :
Date :
Day :

Visitors	Schedule

Problems	Safety Issues

Summary Of Work

Signature : _____

Employee	Trade	Hours	Overtime

Equipment On Site	No. Units

Materials Delivered	No. Units	Equipment Rented	Rate

Others

Notes :

Project Name :

Foreman :

Project No :
Date :
Day :

Visitors

Schedule

Problems

Safety Issues

Summary Of Work

Signature :

Employee	Trade	Hours	Overtime

Equipment On Site	No. Units

Materials Delivered	No. Units	Equipment Rented	Rate

Others

Notes :

Project Name :

Foreman :

Project No :
Date :
Day :

Visitors

Schedule

Problems

Safety Issues

Summary Of Work

Signature :

Employee	Trade	Hours	Overtime

Equipment On Site	No. Units

Materials Delivered	No. Units	Equipment Rented	Rate

Others

Notes :

Project Name : _____

Foreman : _____

Project No :
Date :
Day :

Visitors	Schedule

Problems	Safety Issues

Summary Of Work

Signature : _____

Employee	Trade	Hours	Overtime

Equipment On Site	No. Units

Materials Delivered	No. Units	Equipment Rented	Rate